FORECAST THE WEATHER

Sue Gibbison

CONTENTS

WILD WEATHER

The United States has experienced some of the wildest weather ever recorded. Maybe this is why **predicting** the weather is so fascinating.

What?	How much?	When?	Where?
Most snowfall in a day	76 inches (193 centimeters)	April 14–15, 1921	Silver Lake, Colorado
Largest snowflake	15 inches (38.1 centimeters) wide and 8 inches (20.3 centimeters) thick	January 28, 1887	Fort Keogh, Montana
Strongest wind	165 miles (265.5 kilometers) per hour	October 24–25, 2005 (during Hurricane Wilma)	Florida
Widest tornado	5,250 feet (1,600.2 meters) wide	May 3, 1999	Mulhall, Oklahoma
Most tornadoes in 24 hours	148 tornadoes	April 3–4,1974	Tornado Alley, South and Midwest

Weather **forecasters** predict what the weather will be like. They base their predictions on information they gather. Then they use a mixture of forecasting methods to predict what will happen. You can gather information about the weather and use a mixture of forecasting methods, too.

Hurricane Wilma

The weather changes from day-to-day and from season to seaso but it is the day-to-day differences that are hardest to predict. I order to predict daily changes in the weather, forecasters need know how the weather works.

All the earth's weather happens in the bottom layer of the atmosphere—the troposphere.

Do you know why jet planes fly above the troposphere? It's because pilots know that their passengers will have a smoother ride when they fly above the weather!

The earth's atmosphere

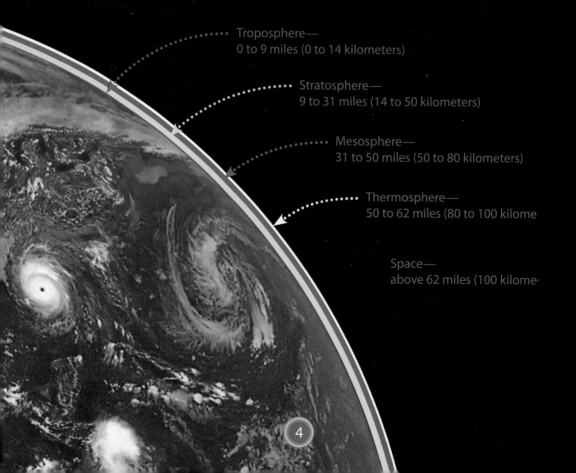

Troposphere—
0 to 9 miles (0 to 14 kilometers)

Stratosphere—
9 to 31 miles (14 to 50 kilometers)

Mesosphere—
31 to 50 miles (50 to 80 kilometers)

Thermosphere—
50 to 62 miles (80 to 100 kilome

Space—
above 62 miles (100 kilome

THE SUN

The sun heats the earth, the water in the oceans and lakes, and the atmosphere. At different times of the year, different parts of our planet are more exposed to the sun. When this happens they warm up.

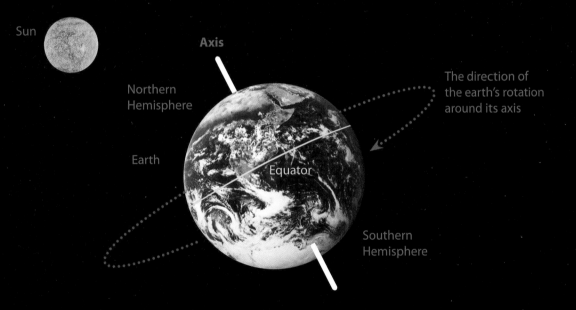

Sun

Axis

Northern Hemisphere

The direction of the earth's rotation around its axis

Earth

Equator

Southern Hemisphere

When the Northern Hemisphere is tilted toward the sun, it has summer and the Southern Hemisphere experiences winter. When the Northern Hemisphere is tilted away from the sun, the seasons are reversed. In between times, the hemispheres have spring and fall.

WIND

As the sun warms air at the equator, this air gets lighter and rises. When warm, **equatorial** air rises and flows toward the North and South Poles, cooler, heavier air flows in under it to take its place. Moving air is what we call wind—and wind doesn't happen just at the equator. Wind direction and wind speed are two of the things that **meteorologists** measure in order to forecast the weather.

Make a Wind Vane

You can find the direction the wind is coming from by licking your finger and holding it up in the air. The side that feels coolest is the direction the wind is coming from. A more exact way of measuring wind direction is to make a wind vane.

You will need:

- two 9 × 12 inch (23 × 30 centimeter) pieces of card stock
- a pair of scissors
- a ballpoint pen that you can unscrew
- tape
- a knitting needle (or chopstick)
- a marker
- a compass

What to do:

1. Using one of the pieces of card stock, cut out an arrow shape like the one shown below.

2. Take the ballpoint pen apart.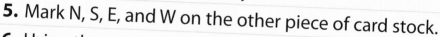

3. Tape the arrow to the top half of the pen, as shown below.

4. Slide the knitting needle where the tip of the pen usually goes, as shown below— the arrow should spin freely.

5. Mark N, S, E, and W on the other piece of card stock.

6. Using the compass, lay the card stock on the ground outside and line it up with north, south, east, and west.

7. Push the knitting needle through the middle into the ground.

8. When the wind blows, the arrow will show the direction the wind is coming from.

Top half of pen

Knitting needle

Make an Anemometer

To measure the wind's speed you'll need an anemometer.

You will need:

- tape • 4 Styrofoam or paper cups • a paper plate • a marker
- a spool of thread • a knitting needle (or chopstick)
- a digital watch (or a watch with a second hand)

What to do:

1. Tape the cups on their sides to the edge of the plate, making sure that they all face the same way, as shown on page 9.
2. Mark a large X on one of the cups.
3. Turn the plate over and tape the spool to the center of the plate.

4. Push the pointed end of the knitting needle into the ground and slide the knitting needle into the spool.

5. Count how many times the marked cup goes around in 1 minute. This measures the wind's speed in **revolutions** per minute.

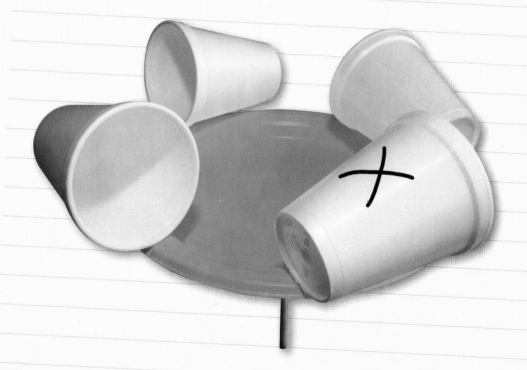

Tell a friend:

Using your wind vane and anemometer, you can now tell a friend the wind's direction and speed.

SPIN

All around the world, warm air is rising and colder air is moving in under it (as wind) to take its place. This is called convection. The rotation of the earth on its axis—its spin— helps to control the direction the wind goes.

A hurricane in the Northern Hemisphere

Did you know that about a thousand tornadoes are reported across the United States in an **average** year.

Hurricanes and tornadoes spin counterclockwise in the Northern Hemisphere and clockwise in the Southern Hemisphere.

A hurricane in the Southern Hemisphere—where they are called cyclones

A Tornado in a Bottle

You can see the effect of spin by modeling a tornado in a bottle.

You will need:

- $\frac{1}{5}$ bottle of vegetable oil (8 ounce or 250 milliliter size)
- a small amount of brown oil paint • 1 teaspoon of ground pepper • 2 plastic bottles that are the same size • water
- tape • a washer that is wider in diameter than the bottle openings • duct tape

What to do:

1. Mix the vegetable oil, paint, and pepper in one bottle.

2. Add water until the bottle is $\frac{4}{5}$ full.

3. Tape the washer to the rim of the other bottle.

4. Turn the empty bottle upside down and tape it to the top of the bottle with the liquid in it. Tape the necks *tightly* together!

5. Flip the bottles over so that the one with the fluid in it is now on top.

6. *Quickly* swirl the bottles several times around to start a **vortex** spinning.

7. Watch what happens.

Show a friend:

To show a friend what happens in the other hemisphere, swirl the bottles the other way!

CLOUDS

Warm air rises because it is not as **dense** as cold air. Because warm air is less dense, there is more space for water molecules in it than there is in cold air. As warm air rises through the troposphere, it cools down. Because cool air is denser, it can't hold as much water. When water in the air cools it **condenses** and turns into clouds. When you see clouds growing bigger, you know that warm air is rising, cooling, and condensing.

Did you know that it takes a million cloud droplets to form one raindrop? It takes billions of droplets to form a cloud. A cloud can weigh millions of pounds (or kilograms).

Cloud formation

Clouds form.

Air cools and condenses.

Warm air rises.

Three Ways Clouds Form

Air cools as it rises to go over the top of mountains.

Air cools when it blows over a colder surface, such as the sea.

The area where warm air and cold air meet is called a front. Warm air cools when it is forced up over cold air along the edge of a front.

Think about what happens when steam touches a cold mirror. The steam forms into tiny droplets of water. Then the drops start joining together. When they're too heavy to stay on the mirror, they run down and drip off the glass—just like raindrops!

Watch carefully the next time you take a shower. After the warm water has been running a while, the air will get steamy. You're making a cloud!

Make a Hygrometer

If you are having a bad hair day it might be because of the humidity—the amount of moisture in the air. You can make a hygrometer to measure the amount of moisture in the air.

You will need:
- tape • a paper or plastic drinking straw • a shoe box
- 1 piece of straight hair at least 8 inches (about 20 centimeters) long • modeling clay • a marker • a folded card
- a blow-dryer

What to do:
1. Tape one end of the straw to the bottom of the shoe box.
2. Tie the hair around the straw, as shown on page 15.
3. Tape the other end of the hair to the top of the box. Make sure the hair holds up the straw.
4. Put a little clay on the end of the straw as a weight to keep the hair tight.
5. The next time you take a hot shower, take your hygrometer with you. Keep the bathroom door and windows shut while you take the shower. If your bathroom has a fan, don't turn it on! Put your hygrometer on the floor. (Don't put it in the shower.)

6. Once the bathroom is really steamy, turn the shower off and mark where the straw points to on the card. This will be close to 100 percent humidity.

7. Open the door and windows and dry the hygrometer with the blow-dryer set on the no-heat setting.

8. Mark where the straw points once you're finished. It will show a lower level of humidity.

Tell your friends:

Put the hygrometer outside, under cover, and watch what happens as the humidity changes from day to day. Use it to show your friends when the humidity is high. Lots of moisture in the air could mean that clouds are forming.

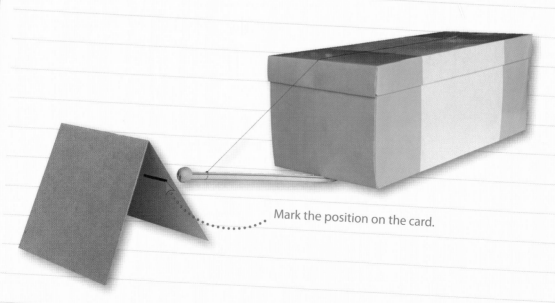

Mark the position on the card.

Using Clouds to Forecast

There are many different kinds of clouds. Meteorologists have given them names, depending on what they look like, where they form, and what they're made of. Studying clouds and knowing their names will help you to forecast what the weather is going to be like. There are three main groups of clouds.

Name	Meaning	Color	Shape	Height	Made of
cirrus	"curled"	white	feathery, wispy	very high in the sky	ice crystals
cumulus	"piled up"	white to dark gray	thick, fluffy	can grow very tall	water droplets
stratus	"layered"	usually gray	flat layers	low in the sky	water droplets and ice crystals

To talk about clouds using the language of weather forecasting, you need two more cloud names:

alto = high, **nimbus** = rain

Meteorologists use different combinations of these five cloud names to give more exact descriptions of cloud types (for example, altocumulus).

Altocumulus clouds

16

Rain, Hail, or Snow?

Clouds tell us about the weather that's coming, but clouds themselves don't always mean rain. It's the color of the clouds that tells us what kind of **precipitation** will fall.

So what's going to come out of that dark cloud that may be above you right now—rain, hail, or snow? That pretty much depends on the temperature. Rain is the most likely result, but if the raindrops get lifted high enough up into the troposphere, they'll turn into ice. Ice crystals form when a cloud's temperature falls below freezing. These crystals fall as snow or hail.

White clouds = nice weather.

Gray clouds or gray patches in white clouds = it might rain, hail, or snow.

Dark gray or black clouds = rain, hail, or snow is not far away!

Hailstones go for a roller-coaster ride around the inside of a cloud. Each time they're carried up, they get a fresh coat of ice—until at last they're too heavy for the air to hold them up any longer. Then they fall.

AIR PRESSURE

There's 9 miles (14 kilometers) of troposphere above you. Beyond the troposphere, the air gets thinner, but there's still a lot of air up there, pressing down around you. If you weighed the air right above you, it would weigh about 2,204 pounds (1 metric ton)! This air is pushing down all around you all the time, but because the liquid and the air in your body is pushing back with the same amount of pressure, you don't feel it.

Make a Barometer

Changes in air pressure are another thing that meteorologists measure to help them forecast the weather. A barometer measures air pressure changes. This is probably a meteorologist's single most important piece of equipment. Here's how to make one.

You will need:

- scissors
- a balloon
- a glass jar without a lid
- a rubber band
- tape
- a drinking straw
- a ruler

What to do:

1. Cut out a piece of the balloon and stretch it over the top of the jar.
2. Fasten it with the rubber band so that it's airtight.
3. Tape one end of the straw to the center of the balloon.
4. Tape the ruler straight up and down on a wall with the bottom of the ruler level with the bottom of the jar.
5. Place the jar so that the straw almost touches the ruler.
6. Check where the straw points to at different times over the next few days. The rubber will bulge up and the straw will dip down when the air pressure falls outside the jar. The opposite will happen when that air pressure rises.

Tell your friends:

Air flows into areas of low pressure to equalize the air pressure. This flowing air (wind) brings different weather with it. If the air pressure starts to change, you can warn your friends that the weather is about to change!

Fronts

A drop in the barometer shows that the air pressure is falling and a front is on the way. A cold front brings cooler air, and a warm front brings warmer air. Fronts often bring rain, hail, or snow. This chart will help you tell whether a warm front or a cold front is approaching.

	Air pressure	Air temperature
As a warm front arrives	falls slowly, as detected by a barometer	warms up slowly
As a cold front arrives	falls quickly, as detected by a barometer	cools quickly

If a cold front is coming, it will move in quickly and force itself under the less dense, warm air around you. There are often showers and thunderstorms ahead of a cold front, but once the front has passed, the storm will be over. Cold fronts create a narrow band of bad weather that moves away quickly.

If a warm front is coming, the warm air will rise slowly up over the cold air. The rain in a warm front stays around longer than it does in a cold front. Warm fronts create a wide area of clouds and rain that moves away slowly.

DIFFERENT WAYS TO FORECAST THE WEATHER

Weather forecasters use a mixture of methods to forecast the weather. So can you!

The Persistence Method

The persistence method says "today equals tomorrow—what the weather is doing now is what it will probably continue to do for a while." Use this method if you are not able to detect any changes.

The Trends Method

The trends method uses mathematics to calculate where weather patterns will be in a few hours or days. For example, if a thunderstorm is 60 miles

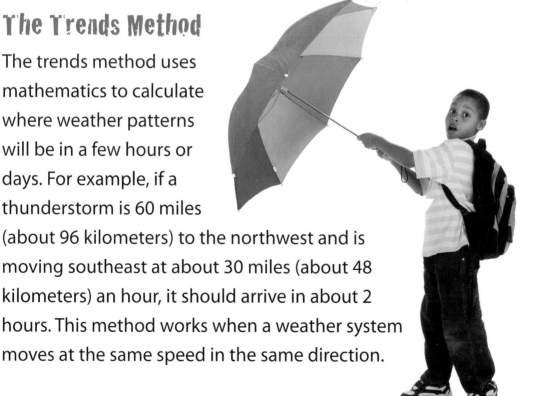

(about 96 kilometers) to the northwest and is moving southeast at about 30 miles (about 48 kilometers) an hour, it should arrive in about 2 hours. This method works when a weather system moves at the same speed in the same direction.

The Climatology Method

The climatology method uses the average temperature and rainfall over many years to predict what the weather "should be like at this time of year".
This method works if the weather is normal for the time of year and you know what it is usually like.

Your Best Estimate

Forecasters use all these methods—and all the information they've gathered— to come up with their best estimate of what the weather will probably be like for the next few days.

So why are forecasts so often wrong? One reason is called the **chaos** theory. This theory says that small changes in the atmosphere right now can make huge differences to the weather in a few days' time. Changes can be so small that they can't be **detected**, even with wind vanes, anemometers, hygrometers, barometers, and other tools, such as weather satellites. The chaos theory is why some scientists say that day-to-day weather forecasts for more than about two weeks ahead will never be possible. At present, five days is about the limit.

People are interested in what the weather is going to do. Most people are fascinated by it. So, what will the weather be like tomorrow? Use the information in this book to help you forecast it!

GLOSSARY

average—the usual or typical

axis—a line through the north and south poles of a globe

chaos—when events are ruled by chance

condenses—changes from a gas into a liquid

dense—compact

detected—identified, noticed, seen

equatorial—from the area around the equator

forecasters—people who predict something

meteorologists—scientists who study weather and climate

precipitation—rain, hail, mist, sleet, or snow

predicting—figuring out what's going to happen before it happens

revolutions—turns

vortex—a whirlpool